YOUR KNOWLEDGE HAS VALUE

- We will publish your bachelor's and master's thesis, essays and papers

- Your own eBook and book - sold worldwide in all relevant shops

- Earn money with each sale

Upload your text at www.GRIN.com and publish for free

Bibliographic information published by the German National Library:

The German National Library lists this publication in the National Bibliography; detailed bibliographic data are available on the Internet at http://dnb.dnb.de .

Imprint:

Copyright © 2017 GRIN Verlag
Print and binding: Books on Demand GmbH, Norderstedt Germany
ISBN: 9783668883093

This book at GRIN:

https://www.grin.com/document/455218

Haider Mallick

Organisational Trends in Groupage Transports Improving Efficiency in Planning and Operations in Pakistan

GRIN Verlag

ORGANISATIONAL TRENDS IN GROUPAGE TRANSPORTS IMPROVING EFFICIENCY IN PLANNING AND OPERATIONS WITH FOCUS TO PAKISTAN

Submitted by: HAIDER JAWED MALLICK

Table of Contents

CHAPTER 1

Pakistan - An undiscovered field for transport and logistics

Pakistan is the 6th largest country in the world by the population (182 million approx) in South Asia occupying one of the key places for the trade routes in warm water connecting Gulf to china in the Indian Ocean. Pakistan is a developing country and is one of the Next Eleven, the eleven countries that, along with the BRICS, have a potential to become one of the world's large economies in the 21st century. According to the latest economic survey, Pakistan has approximately 264,000 km of road network catering to over 90-95% of inland freight and 7800 km of railway tracks.

Fig 1.1 Trucks on the highway in Northern Pakistan

[Image was removed for the publication.]

"The Express Tribune, January 18th, 2016"

The current situation is that Country is democratically governed by a large political party but there is a political instability causing risk of more deteriorating law and order condition. Some parts of the country had suffered and were affected by the ongoing War on terror in Afghanistan, for example, full military control in some areas causes immobilization of people and freight, further unfavorable relations with big neighboring country INDIA is also affecting adversely.

On the other hand, a new project for planning and developments of energy resources as well as CPEC (China-Pakistan Economic Corridor) are progressing. Developments of these projects are extremely expected to bring prosperity and economic boom not only in Pakistan but in the region of south Asia.

Pakistan's logistics mostly rely on the road network. According to the World Bank statistics, 96% [1.1]of the national freight traffic is carried on road networks. The main reason of this is the mismanagement and absence of proper administrative activities in Pakistan Railways' freight system, which has recently started again after the freezing period of more than two years. Due to unbalance market there is an overemphasis on trucking in Pakistan and despite an outdated fleet trucking, it is the backbone of freight transport in Pakistan. In comparison with Europe, time taken for freight journeys via road is twice.

The organizational trends in the LCL(Less container Load or lose container load) and groupage transports of the country are also not fitting in the desired criteria. LCL shipments are mostly used by small and mid-size businesses because generally, they don't have enough volume for full container load (FCL) and due to less lead time waiting is also not possible, hence delivery deadline cannot be missed. Companies and individuals have to go through multiple hurdles during the shipment processes of goods. Special considerations are required specifically at the time of pre-haul and post-haul transports.

In this report, the above few highlighted complications along with others are discussed thoroughly and critically which could be very beneficial for the companies and investors who are planning to Invest in the logistics sector as well as intended to start a business in Pakistan. Referring to these complications my questions for this seminar report are

     i.   What is the prevailing situation of Road transport of Pakistan?

     ii.   Adjacent challenges in the groupage transportation mechanism of the country?

     iii.   How to overcome the discussed challenges and ensure a successful future?

CHAPTER 2

Key Terms, Resources and Methodology for the Paper

2.1 KEYTERMS

Following are the little descriptions of the key terms used in the reports.

<u>Organizational Trends</u>: In this report, it refers to the customs and behaviors practiced in logistics companies of Pakistan.

<u>Supply chain management:</u> The management of the flow of goods and services, involves the movement and storage of raw materials, of work-in-process inventory, and of finished goods from point of origin to point of consumption.

<u>SMEs</u>: Small and medium enterprises

<u>Developing countries</u>: Refers to the countries which are under development

<u>Strategy</u>: A plan of action designed to achieve a long-term or overall aim.

<u>Groupage Transports</u>: It is the consolidation of more than one small consignments to make a complete <u>FTL (Full Truck Load),</u> or also called as <u>LCL(less than container load).</u>

<u>FTL (Full Truck Load):</u> FTL can be also referred as FCL full container load, the consignment of a complete container for some particular customer or destination.

2.2 Resources of Information

Proper information and collection of data include different sources. Due to insufficient proper books, data and information are obtained by internet sites, Newspapers, Magazines, Official Reports, and emails. Telephone interviews from the professionals of the freight industry along with several shippers have also included the sources for this paper.

(Using business strategies models)

2.3 METHODOLOGY

Research methodology is the procedure to find out the systematic solutions for the research problems.

LCL (Loose container Load) or less than truckload shipping is usually concerned with Inland freight as well as with ocean freight but in this paper, the author is more focused on the analysis of Inland LCL (groupage) freight system in Pakistan.

In this paper, research is done by using analytical research method and to some extend the descriptive method to find out the solutions. The application of these methods is by Porters Five Forces Model (comes under).

2.1 Porters 5 forces Model

The methodology used to analyze and describe the different aspects and influencing forces in this report is the Porters 5 forces model, the reason of this is, it helps to figure out the best competitive advantages and possible problems from the aspects of suppliers, buyers, substitutes, new entrants and industry competitors.

The five forces in Porter's model are as follow:

1- Bargaining power of buyers
2- Bargaining power suppliers
3- Threat of new competitors
4- Threat of substitute products
5- Industrial rivalry.

CHAPTER 3

Organizational Trends and Pakistan Logistics: An overview of Groupage transport from Porters Five Forces Model

Transports sector in any country plays an important role to enhance the ability of global competitiveness not only in the economy of the country but it also contributes to the competent functionality of domestic supply chains. In the developing regions, transportation and telecommunication are the two factors, efficient use of which could bring an enormous reduction in the production and transaction costs. Having efficient transport sector can boost the trade competitiveness of any country globally.

In this era of the global village, the importance of cross-border trades and transaction cannot be ignored as the economic growth of any country is highly depended on this and consequently, the importance and need of efficient transportation system emerge. Efficient and proper Transportation System holds the ability to perform a noteworthy role in the economic growth of developing countries.

Strong logistics and transportation services mechanism can enhance the competitiveness of an economy while, inefficient and improper supply chains through high transport and logistics costs could impede not only export and import flows but domestic supply chains which may increase cost for firms, especially those competing in the international market.

The Logistics sector is estimated at 14% of the global GDP (10-30%).Direct transport costs are between 30-40% of all logistics costs; logistics costs are typically 10-30% of final product costs. [3.1]Transportation cost accounts for 39% of the logistics cost (logistics cost accounts for 10% of the sales) [3.2] The transport system has direct and indirect linkage with all the important sectors of the economy. The size of transport infrastructure affects the economic development of any country [3.3]. An efficient and good quality transport system contributes to economic growth by lowering production cost.

It is necessary to understand logistics performance at the country level so that Trade and Transport Facilitation (TTF) system could be evaluated and improved over time and across countries. Lower costs for logistics reduce the cost of delivering products, thus encouraging sales, increasing trade, opening new markets and generally encouraging business. Performance evaluation also helps to improve the efficiency of supply chains and the functioning of related infrastructures, services, procedures and regulations.

The core focus of this paper is to analyze deeply the Groupage freight transport system in Pakistan. The general understanding of the process of groupage transports in developed countries is split into three steps which are Pre-haul, main-haul and post-Haul transports. After describing the little importance of logistics, the groupage freight industry and its system will be observed through Porters Five forces model.

Michael Porter provided a framework model that analyses any industry being influenced by five different competitive fundamental forces. Since this model is launched, many authors used this to analyze different industrial sectors and mostly manufacturing industries. In this report, the same model used to make a thorough analysis for the groupage freight transportation with the context to Pakistan.

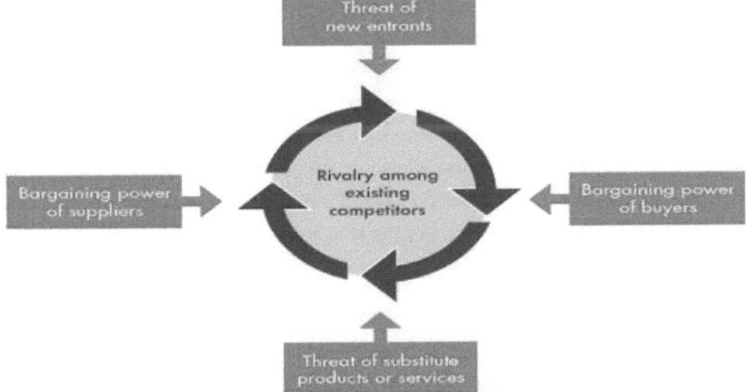

Figure 3.1, porter's framework customized to transport sector Source: - Porter, Michael E. 1979. How competitive forces shape strategy. Harvard business review, 57(2): 137-145.

Before jumping into the Porters five forces model directly it would be helpful to have some idea what are the organizational trends and cultures running in this sector. The freight and transport industry can be counted among the roughest and tough industries in the country. The trucking sector and especially drivers and small companies are in collaboration with one another under a platform called as "ADDAs". These agencies can be referred as "breeding grounds" for the small companies or individual entrepreneurs. In the current scenario, it is somehow essential for these entrepreneurs to join these agencies or "ADDASs" so that they will be the part of these so-called unions in order to be protected by unwilling consequences. Being a part of this type of collaboration could bring many benefits i.e. The ADDA people safeguard and help them out from issues like theft of goods, accidents, street fights over petty issues and they also assist them in the operations of their businesses. On the other hand, you have to deal with some sort of tradeoffs like you would be able and authorize to operate your business services according to the parameters established by the norms of ADDA people. In normal practice, freight forwarders are basically supposed to get the assistance of these ADDAs to complete their transportation as they are the ones who plays the role of main Transporter and carriers. They assign trucks to perform the services for the freight forwarders and logistics service providers. Freight forwarders who do not own vehicles are highly dependent on them.

3.1 Bargaining power of Suppliers

Any type of business requires some sort of inputs like labor, raw material, equipment, services and much more. These types of costs play a significant role in the profitability as well as in the operations of companies. With the aspect of Pakistan, in Groupage freight logistics bargaining power of supplier mostly depends on their economical conditions and business situations. Bigger players are able to secure long-term agreements with their suppliers at much competitive prices. Labors are available at cheap costs in most of the areas due to which loading unloading handling, as well as the assembly and packaging of goods and shipments, are not very much cost sensitive for the forwarders and people carrying forklifts are also available near ports for heavy and big objects. On the other hand, continuous increase in fuel prices is a big challenge being faced by transporters and trucking industry on regular basis.

While talking about LCL transports the importance of warehouses, cross docks and distribution centers cannot be ignored. These places play an essential part in the network of groupage transports. The network of groupage system reflects the "Hub n Spoke" system, which in easy words can describe as "few to many" or "few to few". The response time of the LSP to the traders and businessmen (Industries) especially dealing with perishable goods can face huge difficulties at some point of times, as the accessibility to these distribution centers and regional warehouses are not very convenient. For instance, in the case of heavy rains in the city like Karachi which is the financial hub of the country, is almost likely to jam for several long hours and in many areas, the improper drainage system causes huge troubles for transportation. So the logistics service providers are needed to be very attentive and up to date about the availability of their resources in order to be able to complete their logistical operations.

Due to the limitation of resources in some cases, traders are supposed to inform logistics service providers to manage space in warehouses according to their needs in advance for the expected dates and times.

In normal conditions like excluding seasonal fluctuations and unexpected debacles like strikes and unpleasant law n order situation, the logistical activities can run smoothly and the freight tariff rates are also lower as compare to other countries, hence this could be stated that comparatively, the bargaining power of supplier is low.

3.2 Bargaining power of Buyers

The power of buyers can be described as the effect that your customers have on the profitability of your business. In the groupage transport mechanism, structure systematic line of actions availability and maintenance of proper resources is excessively important. In the scenario of Pakistan, the bargaining power of buyers (users of logistic services) is high. Although the seasonality effect shifts the balance towards the logistics service providers for some duration of the year. The nature and situation of logistics service providers are informal and fragmented and this has a major negative impact on the bargaining power of the suppliers.
The road freight industry is operated by a very large number of small fleet carriers having one or two trucks. Most of the owners are the truck drivers themselves or their friends or family members. Unfortunately, many of them are uneducated, untrained and are performing all activities based on their previous experiences. Besides this, the low switching cost is also a factor that increases buyer's power thus enjoys domination.

The example of the following picture (figure 3.2) a normal day at a central fruit market in a major city of Pakistan, Few trucks standing in a row selling oranges in bulk. In these trucks the oranges being sold by the vendors are not of one person even in one truck, there is a great possibility that this belongs to more than one person or even small companies or retailers. This unorganized system leads to mismanagement and consequently shifts power towards buyers.

Fig 3.2 Fruit market in Lahore

[Image was removed for the publication.]

3.3 Threat of Substitute

In a report published by NUST, Islamabad Pakistan [3.4] road freight system of the country still has no proper substitute. However, substitute for the service is possible among the different modes of freight transport.

    i.   <u>AIR FREIGHT</u>

The Air cargo industry of Pakistan has a rapid development over recent years. ACAAP (Air Cargo Agents Association of Pakistan) is an association of IATA (International Air Transport Association) approved agents. This organization was established in the year 1993 and works under the Ministry of Commerce, Government of Pakistan. Their mission is to facilitate international and domestic trade by providing professional services and to promote harmony among the different airlines, shippers and various government departments involved in Import-Export processes of Air cargo. PIA (Pakistan International Airlines) is the largest and oldest Air Lines of the country. It has faced several ups n downs throughout its journey of passenger and freight transports. Currently, the cargo growth maintained around 1.2 percent declining from 6.4 percent during 2013.[3.6] Other airlines like "Shaheen Air" is domestically involve in air freight transports along with other national and international cargo services like "TCS". Due to high cost and not maximum capacity and operating schedules, it is still crucial for small logistics service provider to offer air cargo facility to their customers at comparatively cheap prices.

ii.   <u>RAIL FREIGHT</u>

The role of Pakistan Railways in the freight haulage is on the decline from the very outset. In the 1950s, its share in freight haulage was over 86% which has now decreased to 4% only [3.4]. Pakistan Railways was the primary mode of transportation in the country till the seventies. However, focusing on road networks and neglecting the importance of railways are the reasons for the decline of the performance and capabilities of the railway system.

Pakistan railways have the network of about 7791 route-kilometers and approximately 589 stations in the network along the country. The railway was built during the British rule on sub-continent and was running successfully. Pakistan railway is more passengers oriented and has not constructed new routes since 1982.[3.5]

Presently railways have not a fixed operation diagram for freight trains. Freight trains are operated in intervals between the running of passenger trains and are frequently forced to stop and wait for the passing and exchanging of passenger trains. In The freight traffic the total number of tons carried and ton-Km have a downward trend since 1965-70 and 1975-80 respectively, and the average distance traveled has been increasing, thereby shifting the Freight traffic from "rail to road" makes sense.

The road freight industry is the predominant mode of freight transportation. It is commonly observed that during the strike of the road freight service providers, the whole economic activity is affected. Hence, this could be stated that substitutes of road freight are not very much adequate.

3.4 Threat of New entrances

This force examines how easy or difficult it is for competitors to join the marketplace in the industry. The easier it is for a competitor to join the marketplace, the greater is the risk of reduction in market share of business. In LCL freight transports system of Pakistan, new competitors could be the new logistics companies or individual truck owners on a small or big scale which may enter the logistics business and also the big international companies which may start their business in the country.

Establishing a new company for logistics services comparatively is not so difficult in the current scenario of the business mechanism. The initial investment required is not too much for starting a logistics Services Company, ownership of one or two trucks and little bit information about transportation and routes of areas in the particular city or country along with personal connections in these ADDA unions enables these individuals' entrepreneurs to join businesses. There is still growth opportunity for small companies in the country but they have to play very strongly in order to capture market share from these old established players and make their access easy to inputs. This is also observed that it is not easy to build strong network collaborations all over the country and to reach economies of scale so that the operational activities could be done effectively and efficiently.

Not only these individuals or small companies are involved in this industry but also some big 3PL (third party logistics) companies are also performing logistics business activities and they are providing full supply Chain management solutions to the traders and customers. For example, AGILITY Logistics and E2E logistics are the ones who provide end to end transportations and logistics services. These types of companies may have a huge room to play in this sector but massive investments are required with strong collaborations with other institutes along the country.

As far as big international giants are concerned, like DHL and FEDEX, they are limited to Courier Express and Parcels (CEP) with full competence but in the sector of freight transports they do not have full operational capabilities and must have to get assistance from local freight forwarders especially in the case of international trades.
In a comparison of both, trucking companies and freight forwarder with (3PL) complete logistics service providers, the threat of new entrance is high in trucking but low in 3PL (third party logistic).

3.5 Rivalry within the industry

Industry rivalry—or rivalry among existing firms—is one of Porter's five forces used to determine the intensity of competition in an industry. Rivalry within the industry in groupage transports is comparatively strong as the service level of most of the companies are same and need to follow the general organizational trend, however, there are only a few companies having 25 to 50 or more vehicles and performing all logistical activities by their own. Tough competition between the small and medium companies makes it difficult to increase market share.

Chapter 4

Planning and Operations Complications in the Groupage Transport System of the country

The groupage transport system in freight forwarding and trucking industry of Pakistan is not completely organized. According to an assumption, there are approximately five hundred small and big companies involve in the business of freight forwarding and transportation across Pakistan. As far as trucking and transportation are concerned, freight forwarders who do not carry their own truck fleet are obliged to get assistance from these truck owner's collaborations to serve their customers requirements. Booking of transportations service, consolidation of goods, loading and unloading of wares, documentations, proceeding to the right destination, deconsolidation of goods and at time delivery of the goods are the some of the issues customers have to be encountered with.

To get the better understanding and knowledge exactly where traders and freight forwarder has to face complications in groupage and LCL transports we have divided the process into three different stages through three different approaches.

- a) Network configuration
- b) Network programming
- c) Network mobilization

Network configuration deals with the allocation and management of resources and inputs. Network programming is concerned with the management of processes and designing the network for the successful and optimal transportation for the customers and Network Mobilization focuses on how to control and implement this designed, defined and suggested network programming.

For the local freight forwarders in the country, availability of resources like warehouses, personnel, and vehicles on the right time at a right place in right quantity is very important. Here some of the main problems being faced by freight forwarders, traders and industries are discussed.

Network configuration

In LCL or groupage transports when logistics service providers are required to perform transportations of shipments or goods for the customers, the exact information about the goods are required like what sort of object is to be transported, what is the volume or weight, how far is the destination, is it within the city or country wide. Based on these typical particulars network configuration is made, like which type of vehicle is required, how many drivers or workers, which type of container or load carrier, would be better. At this level the usual complications they have to face is as under:

Availability of Vehicles
In some cases, this is the first hurdle which the logistics service providers are supposed to resolve. The availability of the best-recommended vehicles for the transportation. If the volume or weight of the goods is enough for the available vehicle or loading unit. In case the required truck is not available and the quantity or weight is exceeding the limit, then they cross limits sometimes.

Fig 4.1 An Overloaded van being fined by Traffic Police in Queta, Pakistan

[Image was removed for the publication.]

Availability of Containers
Availability of containers is the other big issue which the logistics service providers sometimes have to cope with. It is noticed that several times the traders of perishable goods has faced problems due to the shortage of reefer containers at the ports. It is also commonly observed that containers are used in protests by the political parties in processions and by governments as well to block the roads in case of jeopardy or insecurity.

Not enough storage at ports
Insufficiency of storage places at ports is also a critical problem for the freight forwarders to perform their task efficiently and for traders also as this causes a delay in the in the release of their freights and they have to pay extra detention for this.

Increasing Cost of fuels
The continuous increase in the prices of fuel and general price level of commodities is the main driver for the rise of costs for all the activities done at the port and by carriers' i.e handling of freight. As LCL needs more handling of goods and in some cases also needed to use multi-modal transports, this makes LCL more expensive.

Network programming

After the allocation of resources next step is to design the network for the respective transports according to available resources for the customers. Logistic service providers are also expected to design such a program or schedule so that optimum capacity capability and economies-of-scale with smooth interfaces would be possible. For this purpose, they have to develop a throw light at each and every step of the processes to make sure the pro-active detection of possible defects. Here some difficulties are mentioned which sometimes LSP has to go through.

Problems in goods receipt at ports

When the goods are imported by a trader or company, authorized freight forwarders on behalf of them receives goods at ports. In this procedure, they have to counter some problems, as the information flow is not through EDI (electronic data interfaces) but on papers. This situation sometimes leads to misunderstandings or mismanagement.

Port to port service

During the import procedures, some service providers refuse to provide full service to the door of customer or to the warehouse of customers. They just assist their customers in the release of their shipments from ports.s4a3 serv5ce 6

Terminals and Warehouses

For the longer distance freight transports through LCL, the arrangements of vacant spaces at warehouses and terminals for deconsolidation and reconsolidation of the freight along the route is also a difficult task. The main reason for this is the insufficient number of proper warehouses and third-party logistics service providers. Karachi international container terminal is not big enough to meet the demands.

The only Arrival notices

The complexity freight forwarders have to deal with for their customers is, till date, there is proper information system working at ports in the country. As a time till 2017, only two major ports are in operation for the imports and exports a) Karachi port and b) Bin-Qasim port, approximately 40 km from Karachi in the south of Pakistan. Due to unavailability of proper information management the administration of these ports are unable to inform exact time and date of the arrival of ships to the ports, mostly they confirm the date of arrival but not the timings and that too only ½ (one or two)days before the arrival of ship container. IGM (import general manifest) is provided to businessmen from shippers which have to be filed by the carrier on behalf of his customer to the port for the unloading of respective container. In the case of some misunderstanding, they may face serious difficulties.

No check and balance from the government

The traders are often complaining about the behavior of the staff at the port about being unfriendly and taking bribes from the customers. There is no check n balance system from the government for the performance of these employees like KDLB (Karachi dock labor board) for de-stuffing. As well as the logistics companies are also taking advantage of this critical situation and charging extra money from the customers. The LCL and groupage transports are not an easy task for small or medium traders and the complication may get twice if the trader is unaware of the local management system.

Network mobilization

According to LPI (Logistics performance index) of World Bank, Pakistan is ranked 68th with an overall score of 2.92.[4.1] performance of Pakistan on the logistics indicators like the quality of trade and transport infrastructure is not better but worse than that of the other neighboring Asian countries. The controlling and implementation of the plans for the entire process of transport, checking out the constraints in the system, leveling the flows in the efficient and optimized manner can be referred as network mobilization.

Inefficient Trucking Sector

The trucking sector of the country has a disadvantage of the weak regulatory system, consequences of which for example are overloading, poor safety and low quality of service. Pakistan's truck fleet is predominately outdated by several decades and run on underpowered engines that have an impact on the logistic performance (World Bank, 2006). The technology of the entire fleet consists of a rigid suspension of obsolete and old Bedford Trucks out of which: 67% are 2 and 3 axle rigid trucks, 8% articulated, and 25% multi-axle trucks. [4.2]

Prolonged traveling time

Long traveling durations has several side effects, especially in the case of perishable goods; there is a high possibility that goods get damaged. Not only this but in the regular schedule of freight transports along the country time taken for freight journeys are twice than Europe and this is mainly due to an outdated fleet and poor and unreliable infrastructure. Karachi as being the main port is the starting point for most of the freight journeys in Pakistan as well as for the landlocked neighbor Afghanistan. The time duration from Karachi to another large city Peshawar in the north which is about 1300 km driving distance is 3 days which is too much for the current demand situation.

Security Risk

Since last two decades many parts of the world is going through the storm of terrorism and Pakistan is one of the biggest victims of it. Being the alliance of UN and other big countries against war-on-terror has awarded Pakistan several side-effects too and unstable security condition is one of them. Considering north and western provinces of the country, inhabitants of these areas faced a lot of trouble in past years. But since 2014 the internal army operations against terrorism made the situation better and many places of the country which were no-go area are now again peaceful. Unfortunately people there still sometimes have to face critical situations and this has affected the businesses of those areas adversely.
In some places, there is the risk of robbery or theft of the containers and law enforcement agencies are confronting them as well.

Fig 4.2. Domestic transportation of wheat in an overloaded Truck

[Image was removed for the publication.]

CHAPTER 5

A potential line of actions to overcome these challenges and Future of Groupage transport of country.

 Pakistan needs to make sure the remedies of these critical complications in the sector of freight transports because healthy transportations system has a key role in the development of the economy of any country. It is understood that during the past ten years or so a lot of work has been done in the transportation sector whereby cities like Karachi, Lahore, and Islamabad have taken various steps to improve the streets and roads and mitigate the traffic problems including police efficiency. But more needs to be done.

In order to cope-up with complications comprehensively discussed in the previous chapter. There are some simple and easily accessible options; implementation of those in author's opinion could cause huge accomplishments.

Making and implementation of laws

It is very necessary to make certain laws for the freight transport sector of the country and if the laws are already made then strict implementation of them is required so that industry will be secured by many unwilling consequences.
Trucking companies and drivers must be bound to follow the traffic rules. No overloading should be allowed; in the case of being caught the high penalty should be paid. If they do so the need of 100% more trucks will be created to carry the same amount of cargo. Thousands of direct jobs for drivers, helpers, mechanics, and cleaners will also be created. A lot of indirect jobs would also be created.
If trucking companies start to pay full taxes, the government can earn huge revenues and new opportunities for finance jobs can be established.

Maintaining Standards

Adherence to the standards of traffic and freight rules will originate a boom in different industrial sectors. As implementation of these rules and regulations required a dozen of different inputs like police, new technology, vehicle quality inspection tools, vehicle repairing centers, the demand for newly equipped vehicles and much more.

Dislodging of outdated Trucks

Dislodging or removal of several decades' old trucks from the industry will be beneficial in several ways. Most of these trucks are underpowered and have been repaired thousand times in the history. Bringing new fleet is not an easy task but will a great contribution towards prosperous future. These old trucks are causing a continuous damage o the roads and highways. Furthermore, they are also causing accidents and loss of valuable lives of people. Every year thousands of people became disable, gets lifetime injuries, damage to properties is caused by disastrous road accidents.

Training and Education of People

The people involved in trucking and labor are mostly uneducated and untrained people. Mostly truck drivers are involved in this activity since generations and they do work on the basis of their experiences. Most of the times they are driving over-time without rest and extreme weathers conditions. Faults in trucks may cause more delay in the transportation and they are unable to check it by their selves.

CHAPTER 6

Outlook

In writing of this research report, the author tried to gather data for the actual current information of the freight transport mechanism of the country through several different resources. It included the telephonic interviews, internet research based articles, reports from different governmental institutions and universities. There is still need to consider this sector on a larger scale for further developments and to start relative teaching programs in universities.

Talking about the Transport industry of Pakistan, we can clearly say that it is very complex, it is a mix of formal and informal service providers, freight forwarders, traders, suppliers, formal and informal infrastructure. The dynamics of the Pakistani road freight is quite different from rest of the world, unlike most other countries Pakistan's inland freight is highly tilted towards the road (96%) as compared to rail which is the only sole operator of railways in Pakistan due to fundamental inefficiencies.

Trucking sector of the country requires immediate improvements by modernizing and upgrading its fleet and processes. One possible approach is that concerned authorities should provide financial assistance to the Trucking companies in the purchase of new or used but standard trucks. Also, lowering the import duty on trucks will also push local manufacturers into the competition with international standardized manufacturers.

As far as the groupage transports are concerned, as we studied in the report that, present day's freight forwarders have no proper system to manage inefficient fuel consumption, scheduling and back-load management so most of the times they charge customers for two-way (back n forth) and customers have to deal with it.

Although things at the moment are not satisfactory but there is a great opportunity knocking at the door, expected to game-changer for Pakistan named as CPEC (China-Pakistan Economic Corridor).
"The China-Pakistan Economic Corridor is opening attractive avenues for investment emerging from economic cooperation between the two rising powers of Asia! With extensive roads, railways, ports, and energy infrastructure being laid down, businessmen all around Pakistan are finding new opportunities that are worth their money and time."
http://www.cpecinfo.com/businessmen

This is about $50 billion project and is wished to revolutionize the complete transport infrastructure in the country if implemented successfully with full spirit. Besides, there is a strong need of having good relations with all neighbor countries to make cross-border trade easy among the countries in the region.

Pakistan can be benefitted by the huge advantage of its Geo-Strategic location by just making little more emphasis on its trucking and road transport sector so that it connects landlocked central Asian countries with the Arabian Sea in the south and can become an International trade hub in the region. To get into a prosperous future it must have to take over the barriers of complications as prerequisites.

REFERENCES

1- (1.1)Sánchez-Triana, E., Afzal, J., Biller, D., Malik, S. and Sanchez-Triana, E. (2013) Greening growth in Pakistan through transport sector reforms: A strategic environmental, poverty, and social assessment. Washington, DC, United States: World Bank Publications.

2- (3.1) United Nation Economic Commission for Europe, 2009. Joint Trade and Transport Conference on the impact of Globalization on Transport, Logistics and Trade: The UNECE Work

3- (3.2) Logistics Trends, Achieving Supply Chain Integration (2011) Available at http://www.denaliusa.com

4- (3.3)Tsinghua science and technology (2008) Tsinghua Science & Technology, 13(4), p. CO2. doi: 10.1016/s1007-0214(08)70091-9.

5- (3.5)Federal Bureau of Statistics Pakistan Statistical Yearbook 2008 GoP,

6- (3.6)http://finance.gov.pk/survey/chapters/14-transport%20final08.pdf

7- Reporter, T.N. (2013) Freight sector inefficiencies: Pakistan suffering Rs150bn loss. Available at: http://www.dawn.com/news/777117/freight-sector-inefficiencies-pakistan-suffering-rs150bn-loss (Accessed: 24 January 2017)

8- Availableat:https://www.researchgate.net/publication/266469087_Transportation_Problems_in_Developing_Countries_Pakistan_A_Case-in-Point (Accessed: 24 January 2017).

9- Tahir Masood, M., Khan, A. and A. Naqvi, H. (2011) 'Transportation problems in developing countries Pakistan: A case-in-point', International Journal of Business and Management, 6(11).

10- Available at: http://trtapakistan.org/wp-content/uploads/2016/01/Road-freight-transport-sector-and-emerging-competitive-dynamics_final.pdf

11- https://www.adb.org/sites/defaul/lin t/filesked-documents/cps-pak-2015-2019-ssa-05.pdf

12- Abdul Waheed (2000) Growth of roads and road transport in Pakistan. Available at: http://www.pakistaneconomist.com/issue2000/issue45/i&e1.htm

13- Arsalan, Dr.M. (2016) Http://pakstran.Pk/docs/reports/International%20Best%20Practices%20-%20Mudassar%20Arsalan%203.1.3%20A.Pdf. Available at http://pakstran.pk/docs/reports/International%20Best%20Practices%20-%20Mudassar%20Arsalan%203.1.3%20A.pdf

14- maxim, victoria (2015) Less than a container load booking process. Available at: https://www.theseus.fi/bitstream/handle/10024/89415/Victoria%20Maxim_LCL%20booking%20process.pdf?sequence=1

15- Terry, L. (2017) Thinking small: The business case for LCL. Available at: http://www.inboundlogistics.com/cms/article/thinking-small-the-business-case-for-lcl/

16- Available at: http://www.finance.gov.pk/survey/chapters_16/13_Transport.pdf

17- Pakistan statistical yearbook 2011 (2017) Available at http://www.pbs.gov.pk/content/pakistan-statistical-year-book-2011

18- An assessment of opportunities and constraints of logistic Industry in Pakistan, IFC 2009.pdf.

19- Choudhary, M.A., khan, N., Arshad, M. and Abbas, A. (no date) Analyzing Pakistan's Freight Transportation Infrastructure Using Porter's Framework and Forecasting Future Freight Demand Using Time Series Models.

20- WSEAS, Mastorakis, N.E. and Jha, M. (2009) Recent advances in urban planning and transportation: Proceedings of the 2nd WSEAS international conference on urban planning and transportation (UPT '09), Rodos, Greece, July 22-24, 2009. Athens: WSEAS Press.

21- www.unece.org/fileadmin/DAM/trans/doc/2009/itc/Conf_01_pesut.pdf

YOUR KNOWLEDGE HAS VALUE

- We will publish your bachelor's and
  master's thesis, essays and papers

- Your own eBook and book -
  sold worldwide in all relevant shops

- Earn money with each sale

Upload your text at www.GRIN.com
and publish for free